STAY FOCUSED

Based on *Attention!* by Rob Hatch

First published in Great Britain by Practical Inspiration Publishing, 2025

ISBN   978-1-78860-821-3 (paperback)
         978-1-78860-822-0 (ebook)

EU GPSR representative: LOGOS EUROPE, 9 rue Nicolas Poussin, LA ROCHELLE 17000, France Contact @ logoseurope.eu.

Want to bulk-buy copies of this book for your team and colleagues? We can customize the content and co-brand *Stay Focused* to suit your business's needs.

Please email info@practicalinspiration.com for more details.

Please visit robhatch.com/stayfocused to view a short video from the author and download additional resources.

Practical Inspiration
Publishing™

Contents

Series introduction

Welcome to *6-Minute Smarts*!

This is a series of very short books with one simple purpose: to introduce you to ideas that can make life and work better, and to give you time and space to think about how those ideas might apply to *your* life and work.

Each book introduces you to ten powerful ideas, but ideas on their own are useless – that's why each idea is followed by self-coaching questions to help you work out the 'so what?' for you in just six minutes of exploratory writing. What's exploratory writing? It's the kind of writing you do just for yourself, fast and free, without worrying what anyone else thinks. It's not just about getting ideas out of your head and onto paper where you can see them; it's about finding new connections and insights as you write. This is where the magic happens.

Whatever you're facing, there's a *6-Minute Smarts* book just for you. And once you've learned how to coach yourself through a new idea, you'll be smarter for life.

Find out more...

Introduction

Do you value your own time and attention as much as marketers do?

It's no secret that marketers have been vying for our attention for hundreds of years. But the stakes have risen and the capabilities available to them are beyond what many of us imagined.

Tony Fadell, the founder of Nest, was also on the team that created the first iPhone. He admittedly has some regrets about the unintended consequences of his creation.

'I wake up in cold sweats every so often thinking, *what did we bring to the world?*' he says. 'Did we really bring a nuclear bomb with information that can – like we see with fake news – blow up people's brains and reprogram them? Or did we bring light to people who never had information, who can now be empowered?'

The answer, of course, is both. There's no doubt the iPhone has been transformational in its importance. The access to information and knowledge with this technology has empowered and democratized entire populations of individuals.

But spend an evening with a family of four. Listen to parents who openly wrestle with the impact it has on their children as they struggle to set limits. There are full-blown arguments between parents and their children driving a wedge in family relationships.

The irony, of course, is that while parents are trying to figure this out, they are simultaneously interrupting these important conversations with a quick check of their own phones.

We weren't equipped for this. We weren't prepared for just how quickly these powerful tools would capture our attention and, more importantly, our time.

Let me be very clear. I enjoy technology. I enjoy my iPhone. I've written much of this book on my Mac or MacBook, and even, at times, my phone. I used Google docs and other apps.

My children all got their first phones in middle school. That was our rule. And, yes, my wife and I have spent hours talking about how to best handle conversations about how and when they are used.

Introduction

Personally, I land on the side of believing that, for all its shortcomings, for all the things that kept Mr Fadell awake at night, I am grateful for all the transformative power he helped bring to the world.

Distractions are everywhere. This is only one example.

This book wasn't written to tell you to take a digital fast. That's up to you to decide. And that's the point.

The flow of information and noise coming at us is overwhelming.

But we get to choose what we let in and how we direct our response to it.

I think we've lost sight of that a bit. We've welcomed that endless stream of information into our heads. More than that, we actively seek it out. And in doing so, we've lost the space in between.

I believe in our ability to reclaim some of what we've lost. I believe in our ability to choose where we give our attention, for what purpose and to whom we give it.

There is a quote that has been mistakenly attributed to psychiatrist and Holocaust survivor Victor Frankl: 'Between stimulus and response, there is a space. In that space is our power to choose our response. In our response lies our growth and our freedom.'[1]

Our freedom to choose is perhaps the highest form of wealth and power.

When we are able to consciously direct our time and attention to things that matter to us, we are able to transform our lives. We are able to affect the trajectory of our careers, start a business, build connections and deepen our relationships with friends and family.

In a time and culture increasingly burdened with anxiety and stress, the freedom to choose provides respite. These conscious acts of choosing how we focus our attention are how we create space for what truly matters to us.

This is where the power of simple decisions begins.

It's in the narrow spaces between what we see and what we do.

The more we seek that space, however small, the greater our ability to widen it, reclaim our attention and live a life of intention. We get to choose what we focus on, which ultimately determines what we get done, today, tomorrow and over the course of our lives.

Over the next ten chapters (ten days, if you fancy treating this as a mini course), you're going to discover ten key principles of more intentional focus and experiment with using them for yourself.

Let's go!

Day 1
Our distracted world

It's 5:30 am and your alarm goes off.

Of course, it's not an alarm clock. It's a ringtone you carefully selected and scheduled on your phone.

Maybe you hit snooze, but more than likely you turn it off and immediately unlock your phone to check something, though you're not sure what yet.

It could be the score of the game you fell asleep watching the night before, but more than likely you aren't looking for anything in particular, you're just checking.

You open Instagram or Facebook to see what's there, scrolling past several posts and then flip over to email because you just remembered you were waiting on an email from a client.

As you start to scroll through your inbox something else catches your eye. It's an email from your boss asking a question about the presentation you've been working on. It's an easy question to answer, so you sit up in your bed, grab your glasses and shoot off a quick response.

You go back to your inbox, trying to remember what you were looking for in the first place, and you see that your boss replied almost immediately. There's a twinge of guilt that she's already up and working. Of course, she's probably emailing from bed as well. It's what we do.

You read her response and she asks if you can meet today. You quickly check your calendar and see that you can. While you're there, you notice you forgot you had a phone call with a new client later.

You quickly switch back to your email to search for your last communication with the client to 'refresh your memory' about the meeting. Suddenly, a slight panic hits you because you forgot to respond to your boss about the meeting. After shooting her a quick reply that you can meet, you remember that you originally opened your email to look for something else.

Eventually, you find what you were looking for and give it a quick scan. You don't need to reply, but decide to shoot a quick, 'Thanks for this. I'll look it over and get back to you later.'

Your alarm went off at 5:30 am. You've been awake now for seven minutes.

You go back to your boss's email to confirm the meeting time and schedule it in your calendar. When that's done, you breathe a sigh of accomplishment and pop back over to check Facebook again. After all, you need a break from all the work you've done.

You scroll for a few minutes. A friend shared an article that looks interesting and you start to read it. Halfway through that you notice the time. Now you're behind schedule.

A quick shower leads to looking for the pants you were hoping to wear and wondering if you know where your favourite shoes are.

You ultimately decide on another outfit entirely but not before leaving your closet a mess and your dresser half open. Vowing to deal with it later, you realize you're even further behind and start looking for a quick breakfast as you make the coffee, let out the dog and feed her.

In the rush, you start feeling anxious about the day and decide to check your phone again to see if anything else has come in that you need to be 'ready for'.

Can I stop now?

I know this sounds familiar because some version of this plays out each morning in the homes of nearly every working adult I know.

What follows is a deliberate process to understand and address the forces at play in our daily lives that contribute to the noise and distract us from the life and work to which we aspire.

Some of this noise is external. But often the loudest is internal.

As we progress, we will identify opportunities to leverage these forces for our own purposes. We will use the power of simple decisions to reclaim the space between stimulus and response and direct our attention to what matters most.

Red-dot reactions: the stimulus

As my business partner Chris Brogan puts it, 'Email is the perfect delivery method for someone else's agenda.'

But it's not just email. It's *every* notification we receive on our computers, tablets and phones.

The dings and buzzes and banners on our devices have us on a very short leash.

It's gotten to the point that even when nothing happens, we instinctively check our screens for the little red dot telling us someone, somewhere has done something.

Our red-dot reactions have led us to live our lives on constant alert. Our reactions are so swift, so instinctive, we are leaving virtually no space between stimulus and response.

We've accepted the default settings. We've given permission to everyone we know to interrupt us at any moment.

We don't allow ourselves the time to focus on what we intentionally choose for ourselves, to define what we want our lives to look like.

We don't choose how and when we want information delivered or which notifications we want to get through and those we should filter out.

We don't recognize the power we have to direct our time and attention to things that matter.

Of course, we like to blame technology for this. But even with the demands of technology, we can find the space for choice.

This is a call to flip the process, to define yourself and your values.

It's a call to ground yourself in the life you most want to create as a filter for making decisions that align with your vision and values.

It's time to choose your focus – and plan how to sustain it.

So what? Over to you...

1. What notifications or alerts do I allow in by default, and how often do they disrupt my attention?

2. When do I most feel like I'm reacting instead of choosing how I spend my time?

3. What would it look like if I created more
   space between stimulus and response?

Day 2
False choices and fractured systems

We are inundated with a variety of choices each day.

Marketers provide these choices claiming that it's about individualizing our lives. But the sheer volume of options alone, from which coffee to which flavour of pasta sauce in the grocery store, can be downright debilitating.

The irony of having so many options is twofold:

- With so many choices, we're paralyzed and don't make any
- With so many choices, we choose quick and convenient over thoughtful and relevant.

Simply put, more choices often result in poor choices.

It's because we haven't defined what constitutes a good choice for us before we're faced with the prospect of deciding.

We bet against our own interests. We claim we have no time, yet we are more than happy to binge entire seasons of our favourite box set.

Nearly 78% of Americans have little to no savings and live paycheck to paycheck. They carry significant amounts of debt and have no ability to weather an emergency.[2]

Yet, we continually make decisions that perpetuate the cycle.

This is what happens when our decisions are not aligned with our goals.

This is the value of your attention.

More importantly, it is the value of your distraction and overwhelm.

When companies make it easy for you, it's probably a good idea to take a breath, find space and then decide.

Everything in our way

Our lives don't seem to have the rhythm they once did.

There have always been times when life seems to have a cadence we can count on. That cadence gives us a sense of knowing what to expect and helps carry us through our days.

As with anything, sometimes this cadence gets out of sync.

It shows up in seasons. Not necessarily the seasons of nature or even those measured by holidays, but we can identify times in our lives when things just work well.

As much as we are aware of the effect of circumstances and forces at work around us, it's important to look for the clues to our success and what role we have in choosing to set the rhythm.

Boundary lines have blurred such that every aspect of our lives feels cheated and the idea of knowing what to do next feels impossible to truly grasp.

I am convinced that the amount of information we consume each day is shortening our lives. It may be a slow death, but the articles we click to 'stay informed' or for entertainment are sucking up precious hours with not much to show for it.

Add up the time you've spent consuming commentary over the past six months. Is it an hour a day? Two hours? More?

It's not that watching your favourite shows or reading articles your friends shared on social media is inherently bad, if that's what you've decided to do.

But all too often we aren't deciding, we are procrastinating, distracting ourselves or simply allowing our focus to be claimed by someone else.

A lack of focus is a symptom. It's a sign that something else is at play. Sometimes there are legitimate distractions. An illness in the family takes a tremendous toll, for example.

But more often than not, we are simply allowing the noise to shape our time, and we don't have to accept that.

The myth of 'just do it now…'

One common reason why we don't stay focused when distractions come by is that we honestly believe that we can *effectively* allow ourselves to be interrupted and *successfully* return to the task without expending some amount of effort to refocus.

There is a cost to interruptions and your 'let me do that now, before I forget' mantra is hurting you more than it is helping you.

When you allow your attention to be pulled away by a thought, an action or some other interruption, even if it just takes a few minutes, you lose focus and it takes work to get it back.

So what *should* you do with the ideas that pop into your head as you're trying to focus on a project, or the question that lands in your messages and derails your focus?

The three Rs and the blank page

Here's my strategy. It's perfect for those, *I need to remember to..., I have to call..., I have to add this..., That's a great idea for another project, I want to share that with....*

For me, these are the interruptions that pull me into *let me do that now, before I forget.*

Instead of fighting a thought away, I allow it to rise to the surface and capture it. My three Rs are Recognize, Record, Return. Here's how it works.

Start your day or your project work with a blank page and a pen next to you on your desk:

1. As a thought comes into your head like the ones I mentioned above, something with the urgency to act, or something you need to remember, *Recognize* it.
2. *Record* it on the blank page so that you can free your mind from having to remember it later.
3. *Return* to what you were doing.

Simple is powerful.

RIRA

Sometimes my brain interrupts me not with something I *need* or *want* to remember. It's trying to distract me with something else, something easier or more fun, and giving me an excuse to quit. Maybe it starts as a daydream, or a cycle of worry, while I'm engaged in a project, out for a run or in a meeting with someone. Brains are tricky. These thoughts seem to come from anywhere.

It's pulling me away because what I'm trying to do is hard.

When this happens, the method I use is *RIRA*:

- Recognize
- Interdict
- Refocus
- Act

(I use this particularly when I'm swimming. About a quarter of the way into my workout, the voice starts and the end is all I want. My mind fills with all sorts of negative talk and other trickery to make me give up. It tells me I don't have time for this, or I should be back at work. It tells me I'm not good at it, that

I should be better than this by now. It tells me to quit, or hints at how it would be okay if I cut the workout short today. Maybe you relate.)

Recognize

Notice this step is part of both methods. We simply have to recognize our thoughts for what they are. It's incredibly important because it's in that space where a decision can be made. At first, it's a very small space and it's hard to see, but when you do it's time to...

Interdict

Interdiction is more forceful than recording. It's an interruption of the interruption and a complete halting of the thought pattern, which is especially helpful in the case of negative talk or unhelpful worry. You can choose a word or a phrase that works for you. I use *Stop!* as it jars me. It sets the stage for me to take back control in order to...

Refocus

Sometimes, it's as simple as getting my mind back to the work at hand. Sometimes, I have to ask myself a refocusing question: 'What's the most important thing for me to be working on right now?' When I have the answer, that's when it's time to...

Act

In the case of swimming, my act is to execute *the set I'm in*. If I am working on a project, I need to get back to the spot I left.

Many times, in the *Refocus–Act* stage, I get really small on my actions.

Just finish this sentence. Just finish this pool length. And I build from there.

RIRA gets me back in the game quickly. (Interestingly, *rí-rá* is Gaelic for chaos. Draw your own conclusions.)

Both methods – the three Rs and RIRA – serve me well as immediate frameworks for maintaining my focus or helping me to refocus.

The first is gentler. Thoughts and interruptions will happen. Recognizing and recording them in order to return to your work is more fluid.

The second is more direct for a reason. Our minds can wander into unhelpful places and take us off track. And we may need that sharp interdiction to help us return. It's a helpful tool. It may be hard at first, but you'll become better with practice.

So now you have some tools for managing interruptions. Next, we're going to look at how you make intentional choices to set yourself up to focus on what matters. What does it look like to put success in your way?

 So what? Over to you...

1. Do I truly value my own attention and focus? If not, why not?

2. When and how do I tend to consume information as a substitute for taking action?

3. How and when can I try out the three Rs and/or RIRA this week?

Day 3

Put success in your way
(Part 1)

I have a tendency to make things harder than they need to be. I've done it my whole life.

Part of that is my desire to do things the 'right way', which really just means 'the way I imagined someone else would do them'.

But more often than not, I made things harder simply by getting in my own way.

Sometimes it was by putting too many things on my list. Sometimes it was by not asking for help. Sometimes it was by not even acknowledging what I had to do in the first place.

And sometimes, it was simply a matter of setting myself up to fail.

We all want to do meaningful work. We all want to be more productive, and we want to feel like we accomplished something. But for many of us, the daily experience feels like a struggle.

The space between what we intend and what we do is where we lose our way.

We make some progress, then fall off track. We go on a short run of effectiveness and productivity, and then it stops. We have a good morning and a bad afternoon.

The tendency is to view this through the lens of personal discipline.

We say things like, 'I just need to be more disciplined.' But if discipline were the answer, we wouldn't be here.

The idea that discipline alone will deliver the kind of life or work we want is a myth.

Our failure to consistently do what we intend is not a matter of discipline, it's a matter of design.

The most productive, successful and satisfied people I know don't have more discipline than anyone else. They've simply designed their lives in a way that supports the behaviours they want.

They put success in their way.

You don't need to force yourself to do something. You need to make it easier to do the thing you want to do.

This begins with three principles:

1. Willpower is a limited resource.
2. Decisions are distractions.
3. Habits are a powerful force to which we are biologically prone.

Let's look at these principles in turn over this chapter and the next.

Willpower is a limited resource

We act as if we should have an unlimited supply of willpower. We talk about it as though we can develop more of it, like a muscle.

But willpower is not something we can bank on.

One of the biggest mistakes we make is assuming that we can rely on willpower to help us make good choices or stick with our goals.

If you are constantly forced to choose between going for a run and relaxing, more often than not you'll opt for relaxing. If you are trying to avoid sugar but the cookies are in the house, you'll probably eat one. And then another.

The solution is not willpower. The solution is environment.

Set up your space in a way that supports the decisions you've already made.

Don't want to eat cookies? Don't bring them home. Want to work out? Put your gym clothes out the night before. Want to focus at work? Shut off your notifications.

Remove the friction. Remove the distraction. Eliminate the opportunity for willpower to fail you.

Put success in your way.

Decisions are distractions

Deciding requires energy, and energy is finite.

If you're required to make too many decisions during the day, you'll suffer from what researchers call decision fatigue.

We often waste energy debating with ourselves over the same things again and again.

What should I have for lunch? Do I want to go to the gym? Should I call this client now or later?

Each of those moments is a small negotiation, but they add up quickly. And the more of them we have to make, the more tired we get.

The most effective people don't waste time on decisions they've already made.

They eat the same thing for lunch. They schedule their gym sessions. They batch their calls.

And they don't question those decisions every day.

They've designed their day to limit decision fatigue and free up energy for more important work.

When you recognize how frequently you're forced to decide things that could be automated or scheduled, it becomes clear that simplification is a form of power.

 So what? Over to you...

1. Where in my day do I rely most on willpower, and how often does that backfire?

2. What's one decision I make repeatedly that I could automate or simplify?

3. What would it look like to remove friction in one part of my environment this week?

Day 4

Put success in your way (Part 2)

Let's look at the last of the three principles we identified yesterday.

Habits are a powerful force

We're not completely governed by our habits, but we *are* biologically wired to form them.

Habits are our brain's way of saving energy. Once a habit is formed, our brain gets to go into 'energy saver mode'. We don't have to think about brushing our teeth or driving to work, we just do it.

And that's why bad habits are hard to break. The energy-saving mechanism kicks in whether or not the habit serves us.

How can we leverage this biological tendency to our advantage? How can we consistently build habits that serve our needs, making them as simple as brushing our teeth and using them as a means of focusing our energy and attention?

The key to building better habits is to start small and keep it simple, because the smaller the habit, the easier it is to maintain. And once it becomes part of your routine, you can build on it.

If you think about it, there really are only six habit categories overarching almost every aspect of our lives and linking simple new habits to these can make a considerable change in a short time:

1. Start
2. Finish
3. Eat
4. Sleep
5. Move
6. Connect

That's it. This list sums up the majority of most days.

How you start and finish your day are two of the most powerful habits you can cultivate. (And while they're in no particular order, I'd actually put 'finish' ahead of 'start.' If I end my day well, if I go through the steps in shutting down, I am inevitably setting myself up for a great start.) What habits could you build onto these foundations?

- Beginning the day with meditation (Start)?
- Emptying the dishwasher each night (Finish)?
- Drinking water with every meal (Eat)?
- Setting an alarm to remind you it's time for bed (Sleep)?
- Going for a run three times a week (Move)?
- Reading a non-fiction book every month (Connect)?

It's simple, but that doesn't mean it's easy. So put success in your way by making your new habits as easy as possible.

For me, the obstacle to running consistently was not the act of running itself. The challenge lay in all of the steps it took for me to get out the door.

So that's where I focused my attention first. I thought through all of the decision points of the morning. Where in the process of getting up, getting dressed and getting outside did I hit a snag?

For me, that meant I had to deal with all of the preparation the night before. Check the weather. Choose the proper clothing. Lay out the shorts, shirt, socks and shoes (I even untie my running shoes). Make sure the phone, watch and headphones are charging. Schedule the time. Decide on the route and place everything right next to the bed.

Now, I can get up each morning and simply run. Why? Because I don't have to think about it. It's no longer necessary to go through the tedious process of looking for what I need to accomplish the simple act of running.

Apply a system of small, simple cues

The environment you create can be a powerful factor when you're trying to create habits. Consider setting up a visual cue that reminds you of what you're trying to do:

- Stick a checklist or tracker on the fridge
- Set your workout clothes by the door
- Place your book on your pillow to remind you to read
- Put your phone in another room when you want to focus.

You're not trying to create discipline. You're trying to create patterns.

Success is not a single event. It's a set of conditions. When those conditions are met, good things happen.

So what? Over to you...

1. What habit do I want to build that feels too big right now – and what's the smallest step I can take to start it?

2. What visual cue or small environmental change could help support one good habit?

3. How could I track or reinforce one small daily win this week?

Day 5
You are the architect of your system

This is one of the most empowering concepts I've learned: I am the architect of my own system. I am in charge of building the experience I want.

This means that the quality of my day is not just something that happens to me – it's something I can build.

Most people approach productivity as a matter of tactics. They look for the right app or the perfect routine. But what they really need is a system – one they design intentionally.

We spend an inordinate amount of time trying to fix what's broken rather than building around what already works. We try to adopt someone else's productivity method, failing to notice that it doesn't suit the way we live and work.

So here's a radical proposal: instead of asking what's wrong with the way you operate, ask what's right.

And then build from there.

Systems over goals

I've never subscribed to the notion that you should be laser focused on your goals. Instead, I believe you should focus on the actions necessary to achieve them.

A goal, by itself, isn't a system. A system is the process by which goals are achieved. And more importantly, it's the process that continues working after the goal is reached.

Having a goal in sight is clearly important. But understanding *what needs to be done* and *setting up the system to do it* is critical because your goal won't mean anything if you don't do the work. You establish systems, then, to help you make the best decisions in the moment and take the most important actions.

For example, if you want to be healthier, you have to eat food that's better for you. The system you need to build to *put success in your way*, then, first has to support what goes into your shopping cart, because long before it goes into your mouth, you have to purchase the right food. So your system here might be as simple as entering the grocery store with a list and a budget.

The best systems are built around your existing strengths and constraints; they draw on what you already know works for you. They are designed to reduce friction and support behaviour that leads to outcomes you care about.

A well-built system allows you to return to your focus even when life pulls you off track.

That's why flexibility is a feature, not a flaw. Rigid systems snap under pressure; flexible ones adapt.

I've worked with dozens of clients who've tried to adopt new routines only to abandon them the first time something unexpected came up.

So we design their systems differently. We build checkpoints. We simplify inputs. We plan for disruptions.

The goal is not to prevent every distraction. It's to create a structure that helps you return your focus to what matters – quickly, consistently and with less guilt.

Routines and rituals

When it comes to designing systems for better habits, I find clients often talk about both routines and rituals. I don't want us to get hung up on words. Whichever you want to use is fine. That said, let me share with you why and how I think they're different.

I think of a routine as something that is more physical; a physical act conducted by our body. Brushing your teeth each morning is a routine, for example. Getting dressed in the morning is a routine.

Routines have a purpose and can certainly be useful.

Rituals are more solemn. They may have physical elements, but they're also a mental exercise. As such, they provide a platform for engaging in something meaningful – and potentially life changing. They help us tune out the distractions and tune in to the moment and our purpose for it.

Have you ever watched a professional baseball player or a college softball player approach home plate?

Elite players have rituals. They perform actions or a series of actions as they enter the batter's box and face the pitcher.

It's more than just a physical routine; it helps them focus on the moment at hand.

The same is true for a basketball player at the foul line. They may bounce the ball three times, spin it in their hands, bounce it two more times, close their eyes, open them, exhale and shoot.

Interestingly, both examples above occur in relatively small moments of time within the context of what are very long games. They occur individually and call for a different level of engagement and focus.

Rituals don't guarantee success. That would be too easy and we'd probably all use them if that were the case.

What they *do* do is prepare your mind and your body to be in the best alignment to do what you're about to do. That is their purpose.

It's important to note that rituals are *not* the same as the process you use to *put success in your way*. That said, you *can* ritualize the act.

In order to improve my efficiency and my mindset for triathlons, I ritualized the way in which I got on my bike.

Having already done all of the preparation of my equipment, selecting the route, scheduling the time and ensuring everything is ready for me to have a successful ride, my ritual for beginning my ride was also important and very specific.

Every time I went for a training ride on my bike, I set up the equipment exactly as it would be in a transition area.

With everything set and ready, I approached my bike in my bare feet, put on my helmet and buckled it.

I put on my shoes. I put on my sunglasses. I jog my bike to the end of my driveway, jump on, take a breath, hit the timer on my watch and start.

Listing the physical steps or observing the process, it appears to be a routine. The ritual occurs when I use the routine to shift my mindset. Rituals point us to what is important and prepare us to focus on it.

How amazing would it be if your rituals prepared your mind to the point that you felt a sense of mastery even before you performed?

Feeling prepared helps me engage in an activity quickly, feeling confident that the right decision about my time has already been made and I have what I need to get to work.

You are not the problem. Your system – or lack of systems – is the problem. And that can be fixed.

So what? Over to you…

1. When does my day tend to flow smoothly – and what patterns or support make that happen?

2. What's one reliable habit or action I can anchor a new system around?

3. Where could I design more flexibility into my routines instead of trying to be rigid?

Day 6
Small – Big – Small

What's your destination? And, more importantly, what steps are you taking today to get there?

The feeling that we're always behind happens to us all. But the more time we spend thinking about what we did or did not do in the last year, the less time we have for doing what we need to do right now.

We all have areas of our business we neglect. We all get overwhelmed with choices. We all feel some sense of not having done enough or that we should have done X, Y or Z years ago. If only we had.

Start where you are

Start where you are, here at this moment. Your business is what it is. And only the decisions in front of you are going to move you forward.

Please don't worry about what you haven't done or what you may have missed. Just start where you are.

So, what do you want to work on? Where are you starting from?

What do I need to do right now?

This sentence is one of the tools that saves me time and time again. If a flurry of activity or interruptions throws me off my game, I stop, take a breath and ask myself this question.

If another project lands on my desk and my mind starts to wander into everything I have to do, I stop, take a breath and ask myself this question.

When I find myself distracted by worry or anxiety about work or my family, I stop, take a breath and ask myself this question.

In every case, it grounds me. It helps me consider whether or not the urgency I am feeling is warranted. It helps me to put things in their proper place.

When you are clear about what is important to you in your business and personal life, it's easier to view even the most unexpected events in the proper context.

What do you need to do right now?

Your day is your week, is your month, is your year

'Your day is your week, is your month, is your year' is a phrase my business partner Chris Brogan and I use all the time. Truth be told, I'm not sure who came up with it. I'll happily give him the credit.

The purpose of the phrase is twofold.

First, it's intended to help you understand that the actions you take today (and the decisions you make) have the potential to impact your entire year.

Second, it's meant to encourage you to look out to your year and define success.

It's a perfect place to ask the question, 'What does that look like?'

Looking out a year from now is a useful exercise to identify the specific results you are hoping to achieve. Then we can begin to deconstruct things and break them down in order to align our actions to the goal.

Let's take a goal of earning $500,000 in one year. It's this simple: if you can identify which actions you need to take each day to generate $10,000 each week in revenue and you execute them, you will indeed earn $500,000 in a year.

The challenge, of course, is identifying the specific actions.

It's also a challenge to stay committed to executing them day after day, week after week.

If your goal is $500,000 in revenue this year and you knew that meeting with five people a day would accomplish this, it now becomes a matter of setting up your day with the daily goal in mind.

The specific actions you take each day, aligned with your goal, become your year.

Everything is a decision

Look around at your desk. Go ahead. I'll wait.

What do you see? Let me take a guess.

A pile of papers, a few books, a broken pair of headphones, four pens, unopened mail, an old picture, a half-filled notebook, a few random folders, a magazine?

I haven't even mentioned your computer desktop. Open tabs and the 'Updates ready to install' message that you keep ignoring, documents you haven't saved and the myriad little red dots?

I'll spare you the bit about all the unread messages on your phone. And the state of your inbox.

Sound familiar?

Every single thing I listed (and I know I didn't get all of them) requires some amount of our attention and a decision.

Heck, even ignoring each of those items, day after day, involves some level of mental energy.

Because each day your brain sees it, acknowledges it and chooses to do something else.

Think about that for a minute. The dish you walked by earlier. The sock on the floor. The magazine you haven't read since you subscribed. All of it tugs at your attention, even for a split second, and forces you to consider whether to address it or leave it.

Hundreds, if not thousands, of micro-decisions every day.

Are you still wondering why you can't find time for your bigger goals?

Small

Each one of those tiny decisions clutters your brain. They sap your energy and keep you from ever getting to the big, important decisions you need to make to grow your business.

I'm not just talking about messy desk stuff. We continue to find ourselves, week after week, considering the same silly stuff. We think we're acknowledging the decision, but we're not.

We subject ourselves to the same thought loops, never really acknowledging that there's a more effective decision to be made.

Until we face this and fix this, we're effectively keeping ourselves from directing our attention towards the bigger, important stuff.

Big

With the clutter of small decisions out of the way, you can see your way clear to find the big, business-changing opportunities. You may have a glimmer of what they are. Maybe you've written down some big goals somewhere at some point.

When's the last time you saw them so clearly that the next steps were almost obvious?

When was the last time you felt confident that the decisions you were making and the actions you were taking were leading you towards the business- and life-transforming goals you know you're meant to achieve?

Small

With clarity on your big picture, you can begin to frame the most critical actions.

Decisions suddenly don't seem as hard. And the long-term impact you want to have naturally prioritizes your time; it guides and frames each decision and action.

The next step is to follow your plan.

The decisions are made. The actions are clear. Keep it small. Trust that the daily actions you make are serving your bigger goals.

And whenever you get off track, start thinking 'Small – Big – Small'.

How it works at work

An example of this in my work is the project of reviewing our year-end financials.

I do this monthly, but I always want to do a year-end review before we send things off to the accountants.

For some reason, I dread this process. It all seems so large and time-consuming and I can feel the resistance mounting in the face of all that has to be done. It feels like hours of my time and I already know I don't like sitting for hours on long projects.

Instead of staying overwhelmed when faced with a project you have to do but may not be entirely excited about doing, it helps to break it down.

In this situation, the first step of breaking it down meant identifying the tools/items I needed.

First on the list was pulling together all of the monthly profit and loss statements and printing them out. These statements, a highlighter and a pen were going to be the tools I needed to get this done.

A browser window open to our QuickBooks account and a window for our bank account and suddenly everything needed to do the work is all laid out.

This step was important for me to get past feeling overwhelmed at the whole project. That was all I decided I would accomplish on this project that day.

Next I established some simple rules:

- One month at a time
- Most recent to oldest
- 40 minutes at a time.

Suddenly the project isn't massive, just a task that needs to be done for 40 minutes in the morning.

It changes from something I've been procrastinating over to something I am doing. The short time spent setting up everything I needed saved days of delaying. And then, it's done.

I know you have goals and projects. What would it look like to break these down into smaller chunks and call each of those a success?

This is the value of your attention and the power of simple decisions.

It is about establishing what matters and making decisions before you have to.

It's about constructing a framework so you can be productive on your terms and available for the things that matter to you.

So what? Over to you...

1. What do you need to do right now?

2. What's one tiny decision you can take today to reduce your mental clutter?

3. What might be the first small step in a big project you're facing right now?

Day 7

Emotional energy and the cost of decisions

We think of focus and productivity as a matter of time. But more often, it's a matter of energy.

And not just physical energy – emotional energy.

Emotions are part of every decision we make. We're not machines executing a sequence of logic. We're people, with moods and memories and fears and doubts.

When you understand how emotion affects your decisions, you can start designing for it – not against it.

The invisible cost of decision making

We rarely notice how drained we are by the constant stream of choices we face.

What should I work on first? Should I reply to this email now or later? Should I take this call? Should I switch tasks?

These choices don't just cost us time – they burn energy. And when energy runs low, emotion takes over.

We start reacting instead of choosing. We seek comfort over progress. We chase distraction, not focus.

And then we blame ourselves for being lazy or unfocused – when in fact we were just depleted.

Design reduces emotional toll

A well-designed system doesn't just save time – it protects your energy.

Think about the places where you tend to lose focus. For example:

- Midday, when your energy dips
- After a difficult conversation
- When a big task feels overwhelming.

These are emotional triggers, not just logistical ones.

When you start to recognize them, you can build support into your day:

- Schedule breaks proactively
- Break big tasks into smaller, less emotionally daunting steps
- Create routines that make decisions in advance – when your energy is still strong.

Emotions aren't the enemy

This isn't about removing emotions. It's about acknowledging them and designing for them. Sometimes you'll feel tired, resistant or overwhelmed. That's human.

But when your system accounts for those feelings – when it gives you a way to keep moving anyway – you stay in motion. You stay focused.

And remember too that when you look back on your most significant personal and professional accomplishments, you'll almost certainly notice they are littered with emotional decisions and actions.

Using emotions to make decisions need not mean being rash or impulsive. Emotions are what drives us to change, and they sustain us through the hardest moments.

When we face a difficult choice, our emotions connect us to the meaning of what we're about to do.

So what? Over to you...

1. What time of day do I feel most emotionally drained – and what tends to trigger it?

2. Where in my work do I second-guess myself most often?

3. What simple structure or routine could help
   me reduce those points of decision fatigue?

Day 8

Decide before you have to

One of the smartest ways to protect your attention is to decide things before you need to.

We tend to make our worst decisions in the moment – when we're tired, rushed or stressed.

But when we pre-decide – when we determine in advance how we'll respond to common situations – we free up enormous mental space.

We also protect ourselves from emotional hijacking.

Pre-deciding beats reacting

I work to eliminate as many decisions from my life as possible. All those tiny little details about what to wear and what to eat get in the way of the bigger stuff. Deciding ahead of time, making sure everything I need is ready and available creates a much easier morning.

The same happens at work. Do I answer this email now? Do I check Slack for new messages? Should I push back on this request?

If you wait until the moment of action to decide, you'll default to the easiest or most emotionally comfortable choice.

Which often means avoiding, deflecting or delaying.

But if you decide in advance – if you create a rule or default – you remove the debate.

You act, instead of agonize.

Decide once

The rule is simple: decide once.

If you've decided you won't check your email until 11 am, you don't need to re-decide it every morning.

If you've decided that Monday mornings are for planning and not meetings, you don't have to justify it each week.

The fewer decisions you leave for the moment, the more freedom and energy you reclaim.

Use the tools you have

This doesn't need to involve fancy new technology; the tools you use every day can become your friends as you design more intentionality and focus into your day.

Your email has filters that can deliver messages into specific folders. By taking a bit of time to set them up, you can get exactly what you want in each folder.

Your calendar allows you to be more deliberate with how you use your time, even your free time. (I sometimes schedule a nap or a walk in the middle of the day – it's far better than filling unscheduled time with random internet browsing and scanning social media.)

Your calendar also has sections for location, contact information and notes about your meeting. Filling in this information means no searching around your inbox for emails or phone numbers two minutes before your next call.

It'll take a little work upfront but imagine what you'd want an assistant to have ready for you. And then become that assistant to yourself. What information would be most helpful to you the moment you need it? What would your day be like if everything was laid out? I'm just guessing, but I'm betting it would help you focus your attention on what matters most.

Define what 'enough' looks like

We're not just bad at reacting in the moment – we're also bad at knowing when to stop.

We keep working late because it's 'just one more thing'. We keep reading because 'what if there's something important?' We keep checking because 'what if they need me?'

The solution is to define 'enough'.

How much time is enough for email?

How many client calls per week is enough?

What does a productive day look like – not in theory, but in practice?

If you define 'enough', you can shut the door and move on – without guilt.

So what? Over to you…

1. How could I be my own most useful assistant?

2. Where do I most often struggle to define 'enough' – and how could I set that boundary clearly?

3. How could I structure my week so that fewer decisions are left to the moment?

Day 9

Narrowing your focus

There's a concept I use with my coaching clients called 'one number'.

It's deceptively simple – and incredibly powerful.

The idea is this: if you had to focus on just one number to guide your actions this week, what would it be?

Not 20 metrics. Not a dashboard full of data.

Just one number.

Why one number works

We are overwhelmed by data. Email counts, social followers, hours billed, leads generated, money made, pages written.

Every metric is vying for our attention.

But more metrics don't create clarity – they create confusion.

Focusing on one number forces you to prioritize. It makes you define what matters most right now.

That number becomes a filter. It helps you decide where to spend your energy – and where to stop wasting it.

Choosing your number

Your *one number* is any measurable number that drives an intended goal-related outcome and increases the likelihood of the success of that outcome, and that depends on context.

- If your goal is to get 30 people to sign up to your workshop, maybe your one number is ten calls a day. You make ten sales calls to get people in those seats.

Be aware: it's likely you won't pick your *one number* correctly right off the bat. We often accidentally pick the goal measurement instead of the number we need to reach it.

Remember: the goal is the finish line. Your revenue target is a goal. Pounds lost or gained is a goal. Book published is a goal.

Your *one number* is the way you get there. Fifty courses sold. Two miles walked. Five hundred words a day.

The real beauty of *one number* is its simplicity. It narrows your focus and aligns your efforts, making it easier to know you're on the right path (or not).

One number, one week

You don't have to pick the same number forever. In fact, the best use of this concept is weekly.

Each week, ask, 'What's the most important outcome I can move forward this week?' Then choose the one number that reflects it. Track that number. Stay close to it. Let it shape your daily actions.

When the week ends, reflect: *Did this number help me focus? Did it align with what matters?*

Then choose a new number for the next week.

Measure. Dump. Refine.

Another helpful way to think about what to focus on is the measure–dump–refine framework: measure results, dump what's not working and refine the method.

Too often we make changes mid-stream without fully understanding the effect or impact. We make adjustments to systems because they don't *seem* to fit, but honestly what are we measuring? How are we using that information for evaluation? How do we know what to cut and what to add?

Measure

It starts with an honest look at what's working and what's not – choosing your metrics intentionally, based on what you most value. Be open-minded at this stage; all you're doing is collecting data. Then you can sit down with the data and see what it tells you.

Dump

Armed with that information, you can see what's working and what's not. Which means you can decide what you're going to stop doing, because it's just not bringing you the results. (Focus is about what you DON'T do, just as much as what you DO.)

Refine

Rather than making ad hoc tweaks to your systems, make intentional changes based on what you've identified through the measuring and dumping phases. Deciding what you're going to change at this stage means you give your systems time to play out and see what needs to change.

Another name for this process is reflective practice, and we're going to come on to that in the next chapter.

So what? Over to you...

1. What's one number I could track this week that would reflect real progress?

2. Is that number something I control – or just a result I hope for?

3. How will I use that number to guide my daily actions?

Day 10

Accountability, integration and momentum

You've made decisions. You've clarified what matters. You've designed systems and reclaimed space.

Now comes the part that ties it together: *accountability*.

Why accountability matters

We don't just need reminders – we need reinforcement.

Most of us know what to do. We just struggle to do it consistently.

That's where accountability helps.

It provides structure. It creates reflection. It increases follow-through.

And it doesn't have to mean a coach or a manager. It can be a simple habit of self-check-in.

Reflective practice

One of the most powerful tools for reclaiming your focus is reflective practice.

There are a number of definitions for reflective practice – my favourite is this one:

> In reflective practice, practitioners engage in a continuous cycle of self-observation and self-evaluation in order to understand their own actions and the reactions they prompt in themselves and in learners (Brookfield, 1995; Thiel, 1999). The goal is not necessarily to address a specific problem or question defined at the outset, as in practitioner research, but to observe and refine practice in general on an ongoing basis.[3]

I was particularly drawn to a couple of things in this definition.

Continuous cycle

Taking time to review your work and reflect on it has to be a regular occurrence – a consistent practice.

Not specific

The 'goal' is 'not to address a specific problem' but to 'observe and refine practice'. For our purposes, this works. You want to get better. I see it as an important tool in the process of directing your attention to the work that matters.

The *challenge* though is that it's one thing to accept this as a premise, it's quite another to be able to put this into practice.

Remember my favourite question, 'What does that look like?'

What does reflective practice look like for you?

We generally focus on two questions: 'What did I do wrong?' and 'What did I do well?'

What's more, we tend to focus on what went *wrong* disproportionately. We believe, and we are not entirely wrong, that if we focus on shoring up our weaknesses, fixing what is broken, we will improve.

As I said, it's not wrong to think this way. Our weaknesses deserve attention. Our mistakes require adjustment.

However, the answers we are seeking to get better rarely lie in what went wrong or what isn't working. They lie in asking, 'What went well?' and, more importantly, 'Why?' and 'How do I replicate it?'

We need to move from exhaustively examining our failures to recognizing them and moving through them. We need to focus our efforts on what went well, to recognize what works in order to replicate it in the future.

In order to improve, we must ask the questions that help us become better.

In order to ask the right questions, we need to allow for the time and space.

Implementing a habit of reflective practice, where we can examine our performance, is critical to growth, and focusing our attention (and questions) on our successes provides us with a framework for replicating success in other areas.

The work is the practice

Athletes go to work every day. They show up at a practice facility, pool or gym every day to do the work that enables them to perform when they are called to. There is a framework that they or their coaches have built for them. They implement routines they know will help them to be successful.

As they go through each routine, working within the frame, they are also making mental notes about what's working, what isn't and what needs to be done differently.

Their work is their practice. The routines they do each day are part of the 'job' of being an athlete. Throughout their workday, notes are taken, ideas come and they are reflecting in the moment about what they need to do differently next time.

At some point, the notes have to be reviewed, by them, by their coaches, and adjustments need to be made. New plans are drawn up for the next day.

Does this sound familiar?

Our focus is on success, on what we are doing well. While we don't bury our heads in the sand about our failures, we focus on building on what works because it gives us a certain momentum. Just asking yourself the question 'What worked for me before?' is a form of reflective practice.

So be encouraged: you're already doing reflective practice. As you make it more intentional, keep it simple.

Give yourself time to review and prepare at the end of the day. I promise that once you do small bits of this, you'll want to find ways to do it more and make a bigger effort of it.

Here are a few useful questions to keep in mind when coaching yourself through this process:

- What would that look like?
- Why did that work?
- What got in my way?
- What is the first thing I need to do to make this work?

Ask the questions. Listen for the answers.

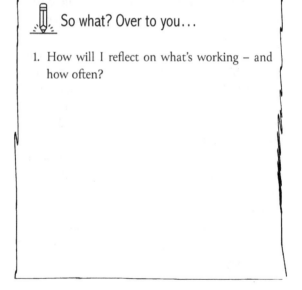

So what? Over to you...

1. How will I reflect on what's working – and how often?

2. What's one small practice I can commit to weekly to reinforce my systems?

3. Who or what helps hold me accountable in a way that works for me?

Conclusion: What does that look like?

What does it look like when things are going well?

This question has become a cornerstone of my work – with clients, with teams and with myself.

It's deceptively simple. But answering it helps you cut through noise, resistance and uncertainty.

It brings you back to a state of intentional living.

This isn't theory – it's practice

Everything in this book is rooted in practice.

Simple decisions. Clear systems. Deliberate structure. Conscious choices.

Not grand reinventions. Not overwhelming change. Just practical steps you can take to regain your time, energy and focus.

Don't aim for perfect

You will still be distracted, lose focus. You'll still get pulled off course. You're human.

The point isn't to avoid all disruption – it's to create systems that help you return to what matters.

Don't wait for ideal conditions. Start where you are. Start small.

And start now.

Keep asking the question

What does it look like when things are going well?

When you answer that honestly – and keep answering it each week, each season, each shift in your life – you begin to write a new script.

You begin to build your days around what you value.

That's the work. That's the opportunity.

Let's stay focused.

(Want more? Visit robhatch.com/stayfocused to view a short video from me and download additional resources.)

Endnotes

[1] Unattributable. Available from https://quoteinvestigator.com/2018/02/18/response/ (accessed 31 July 2025)

[2] CareerBuilder (2017) 'Living paycheck to paycheck is a way of life for majority of U.S. workers, according to new CareerBuilder survey' (2017). Available from http://press.careerbuilder.com/2017-08-24-Living-Paycheck-to-Paycheck-is-a-Way-of-Life-for-Majority-of-U-S-Workers-According-to-New-CareerBuilder-Survey (accessed 31 July 2025)

[3] MaryAnn Cunningham Florez 'Reflective teaching practice in adult ESL settings' in *ERIC Digest*, ERIC Development Team (2001). Available from https://files.eric.ed.gov/fulltext/ED451733.pdf (accessed 31 July 2025)

Enjoyed this?
Then you'll love...

Attention! The power of simple decisions in a distracted world by Rob Hatch

In a world of endless distraction, we have given away two of our most valuable assets: time and attention.

Technology has given us the incredible gift of access to an ever-increasing amount of information and has opened the door to a vast array of choices and opportunities.

However, having more options doesn't correlate to an increase in our success. Research shows that having more to choose from causes anxiety and decreases our likelihood of taking action. We have become paralyzed and polarized, reacting instead of acting and ceding control of our decisions to a continuous onslaught of information, marketing and interruption.

We live in an age where we struggle to decide which information is real or fake. We find it challenging to make even the most straightforward decisions for our happiness and success in our lives and business.

This book will help you reframe your relationship with the demands on your time, overcome decision fatigue and understand the value of creating space.

Rob Hatch sets out a powerful framework and flexible approach that gives you the space to focus your attention on what is important, the power to make decisions aligned with your goals, and the ability to take action with confidence.

Other 6-Minute Smarts titles

 Building Great Teams (based on *Workshop Culture* by Alison Coward)

 Collaborate Better (based on *Collabor(h)ate* by Deb Mashek PhD)

 Customer Success Essentials (based on *The Customer Success Pioneer* by Kellie Lucas)

 Do Change Better (based on *How to be a Change Superhero* by Lucinda Carney)

 Find Your Confidence (based on *Coach Yourself Confident* by Julie Smith)

 Get That Promotion (based on *Getting On* by Joanna Gaudoin)

How to be Happy at Work (based on *My Job Isn't Working!* by Michael Brown)

How to Get to Know Your Customer (based on *Do Penguins Eat Peaches?* by Katie Tucker)

The Listening Leader (based on *The Listening Shift* by Janie Van Hool)

Managing Big Teams (based on *Big Teams* by Tony Llewellyn)

Mastering People Management (based on *Mission: To Manage* by Marianne Page)

No-Nonsense PR (based on *Hype Yourself* by Lucy Werner)

Present Like a Pro (based on *Executive Presentations* by Jacqui Harper)

Reimagine Your Career (based on *Work/Life Flywheel* by Ollie Henderson)

 Sales Made Simple (based on *More Sales Please* by Sara Nasser Dalrymple)

 The Speed Storytelling Toolkit (based on *Exposure* by Felicity Cowie)

 Write to Think (based on *Exploratory Writing* by Alison Jones)

Look out for more titles coming soon! Visit www.practicalinspiration.com for all our latest titles.

www.ingramcontent.com/pod-product-compliance
Lightning Source LLC
Chambersburg PA
CBHW031225050326
40689CB00009B/1485